PRIME NUMBERS'
CHARACTERISTICS

WHY THEY ARE WHAT THEY ARE

L . J. Balasundaram

Harvard Institute for Learning in Retirement

Cambridge, Massachusetts

ISBN-10: 146378869X
ISBN-13: 9781463788698

Dedicated To
the Memory of My Parents

PREFACE

Prime numbers have been the subject of much study and commentary; Leonhard Euler, Gauss, Hardy, Ramanujan, Erdos, and Pomerance have all contributed to the great strides made in exploring and extending the vast domain of prime numbers. Still, some aspects of prime numbers continue to be shrouded in mystery, and this makes them all the more interesting and intriguing to its devotees. This book is an attempt to goad the prime number deity into revealing some of its esoteric secrets.

One of the many tantalizing unelucidated issues associated with prime numbers is their distribution among natural numbers. Though there appears to be some kind of long-range sequence and some sort of rule(s) governing prime formation, the lack of order offers fertile ground for those engaged in prime number investigation. A search for the controlling factors that govern prime distribution prompted the author with prime number investigation; to his surprise, the author found very simple rules governing them. Such rules are already known in some fashion in the realm of prime numbers and these are applied to resolve some of the many un-resolved issues associated with prime numbers.

During the course of my investigation, I consulted with many colleagues and read references too numerous to list; I have listed only those references that have a direct bearing on the text. I have reproduced Fig. 5.2 of The Book of Numbers to illustrate the accuracy of my computation of $\pi(N)$. I am grateful to Professor Fran Schied for helpful discussions, Dr.Robert Lurie for suggesting solutions to computer problems, Dr. Sydney Diamond for encouraging discussions, Dr. Tim Lynch of Psychsoft, Inc. for ungrudging computer consultancy, and above all, to my wife Gul, for help and support. The views advanced in this book are mine and do not reflect those with whom I had discussions or consultations.

March 2009   L. J. Balasundaram

PREFACE TO SECOND EDITION

The book's title is revised to "Prime Number Characteristics - Why They Are What They Are", from "Prime
Numbers – Some Characteristics" to reflect book's contents. In addition, to correcting errors, Chapter 4 title
and contents and Chapter 5 title have been revised.

November 2011   L. J. Balasundaram

CONTENTS

CHAPTER 1

NUMBERS AND THEIR REPRESENTATION

INTRODUCTION

It is not definitely known when humans learned to count, but it is probably a safe guess that the advent of counting is related to the ability of humans to touch the thumb with fingers. Finger touching probably led to counting, which in turn led to a counting system that in turn led to evolution of numbering systems. Humans living in different parts of the world evolved differing number systems, but the system commonly in use today, is the Hindu-Arabic system of numerals developed by Hindus in India and passed on western world by trading Arabs.

As civilization advanced, the numbering system diversified to meet different uses the system was put into that ranged from prime numbers by Eratothenes to Pythagorean theorem. This necessitated diversification to real and imaginary numbers. Real numbers were further divided into natural, whole, integers, rational, and irrational numbers.

Natural numbers are positive integers arranged in ascending order starting with 1 and represented as 1, 2, 3, 4, 5, ... They represent an arithmetic progression with a common difference of 1. To classify prime numbers, one has to start with its definition of as a positive integer greater than 1 and having as its divisors 1 or itself. This definition dictates that first even and odd numbers greater than 1, viz. integers 2 and 3 meet criterion of prime number definition and therefore are primes. Positive integers greater than 3, in order to meet criterion of prime number definition, must not be divisible by 2 and/or 3 and therefore have to take the form 2*3*NA - 1 or 2*3*NB + 1 with NA and NB taking up positive integer values greater than 0. All prime numbers greater than 3 must therefore of the type 6*NA -1 or 6*NB + 1, and an examination of prime number listings confirm this to be the case. Prime numbers may therefore be classified in two groups – Primary primes and Derived primes. Integers 2 and 3 are classified as Primary primes (or first generation primes) and all positive integers greater than 3 and of the type 6*NA - 1 and 6*NB + 1 are classified as Derived primes (or second generation primes) because their forms are derived from primary primes 2 and 3.

Even though representations 6*NA – 1 and 6*NB + 1 are used to describe derived primes, prime numbers are not generated for all positive integer values of NA and NB when substituted in these representations. This is because primes generated by these representations when multiplied with each other generate integers which in turn can be represented by 6*NA − 1 and 6*NB + 1 representations resulting in primes and factorable integers appear in 6*NA -1 and 6*NB + 1 representations. Such factorable integers are termed "Compound Primes". This terminology is used to distinguish this group of factorable integers from composites whose factors include 2 and/or 3. It must be emphasized the term composite as defined here has a different connotation from that used in conventional mathematical texts. Conventional definition of term composite includes all factorable integers, but as defined here it only refers to a group of factorable integers that include only 2 and/or 3 as factors. Armed with these definitions, natural numbers can be represented pictorially in Fig. 1

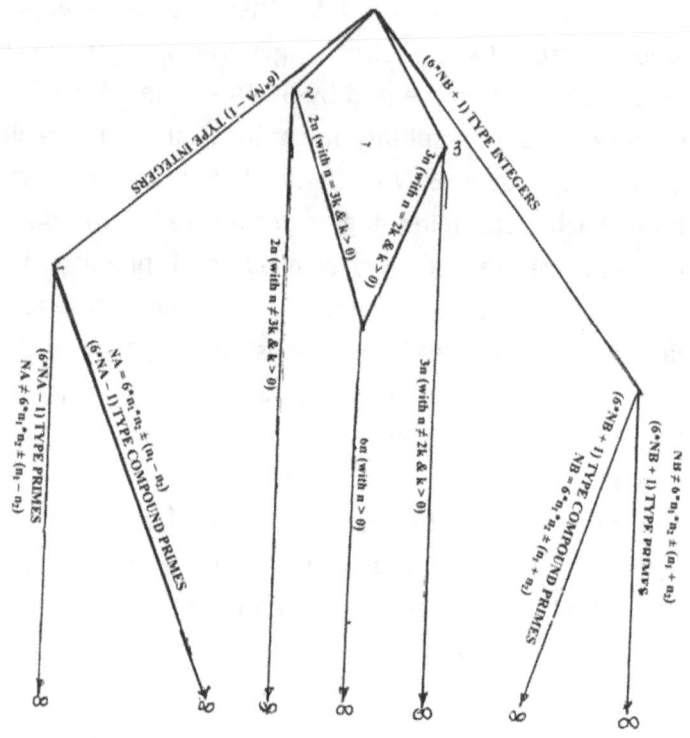

PRIME: A POSITIVE INTEGER GREATER THAN 1 WITH DIVISORS 1 OR ITSELF.

COMPOSITE: A POSITIVE INTEGER WITH 2,3 OR 2 AND 3 AS FACTORS.

COMPOUND PRIME: A FACTORABLE POSITIVE INTEGER GREATRER THAN 3
WITH PRIMES GREATER THAN 3 AS FACTORS.

NATURAL NUMBERS -- A SCHEMATIC REPRESENTATION

To better understand some of the properties of prime numbers, it is proposed to address them using the following framework:

1. Why is there no general formula for generating all prime numbers?.

2. Why do some odd positive integers meet the requirements of the definition of prime numbers and others do not e.g. 71 and 77?.

3. Presence and absence of twin primes in sequence of prime numbers

4. Why is $\pi(N)$ for any N a certain number and not any other number. i.e. why is $\pi(100)$ is 25 and not 32 or 47?.

5. Why does an even number like 198 have 13 Goldbach prime pairs (i.e. pairs of primes meeting the requirements of Goldbach Conjecture which is defined as any positive even number greater than 2 can be expressed as the sum of two primes), but adjacent even numbers 196 and 200 have only 8 and 7 Goldbach prime pairs respectively?.

To address the above questions, the following terminology is proposed. Some symbols and terms given below are those used in conventional mathematical literature, while the terms in bold face are defined to convey intended ideas.

Prime(s): Positive integer(s) greater than 1 with only 1 or itself as divisors.

Primary Prime(s) or First generation Prime(s): Refers Primes 2 and 3.

Derived Prime(s) or Second generation Prime(s): Primes other than 2 and 3.

Composite number(s): A factorable positive integer(s) whose factor include 2 and/or 3.

Semiprime(s): A natural number that is the product of two prime numbers.

Compound prime(s): A factorable positive integer whose factors are only derived primes and do not include integers 2 and/or 3.

N, N_1, N_2, n_1, n_2: Positive integers.

NA, NB: Derived primes/Compound Prime groups/ numbers.

P_{na}, P_{nb}: Total number of prime and Compound prime sites of NA and NB groups respectively in the group equal to or less than N.

π (N): Number of primes that are equal to or less than N.

NSP: Number of sites for occupation by derived primes and compound primes less than N of an even number 2*N.

Plh: Number of derived primes greater than 4 and less than N of even number 2*N.

Clh: Number of compound primes greater than 4 and less than N of even number 2*N.

Puh: Number of derived primes greater than N but less than (2*N -3) of even number 2*N.

Cuh: Number of compound primes greater than N but less than (2*N -3) of even number 2*N.

Np: A representation of prime characteristics of integer N of even number 2*N. If N is a prime, Np =1, If N is not a prime Np =0.

CHAPTER 2

A SIEVE FOR PRIME NUMBERS (GREATER THAN 3) GENERATION

Over the years, many prime-producing formulas[1] have been proposed, but none have been able to generate **all** primes. Eratosthenes[1], the famed librarian of Alexandria, devised a sieve to generate prime numbers by filtering out factorable integers from a natural number list. Extending this sieve concept, a criterion for generating all primes greater than 3 is proposed. As compound primes are generated by multiplication of derived primes of form $6*NA - 1$ and $6*NB + 1$ patterns, the NA and NB values at which they appear in the two patterns may be written as:

$$NA = 6*n_1*n_2 \pm (n_1 - n_2) \ldots\ldots\ldots \text{Equation A.}$$

and

$$NB = 6*n_1*n_2 \pm (n_1 + n_2) \ldots\ldots\ldots \text{Equation B.}$$

where n_1 and n_2 are positive integers taking up values greater than 0. Filtering out each set of NA and NB values generated by equations A and B above for all integer value of n_1 and n_2, at which compound primes appear from natural number lists leaves two residual number lists for NA and NB. Substituting the residual list number of NA in 6*NA − 1 and residual list of NB in 6*B + 1 will generate all primes greater than 3 and is in exact agreement with those reported in literature[2]. Thus we have a framework – albeit indirect – to generate all primes greater than 3.

$$f(N) = (6*NA - 1) \text{ where } NA \neq 6*n_1*n_2 \pm (n_1 - n_2)$$

and

$$f(N) = (6*NB + 1) \text{ where } NB \neq 6*n_1*n_2 \pm (n_1 + n_2)$$

where n_1 and n_2 are positive integers taking up values greater than 0.

It is no surprise, therefore, that no single simple formula can exist to generate all prime Numbers.

Table 1 lists primes 5 to 1000 generated by this approach.

Using the same approach, and counting the number of generated primes, while Table 2 shows there are 25 primes between 100 digit number interval 0 and 100 , but there are only 2 primes in the 100 digit number interval between 10,000,000 and 10,000,100[2].

One can surmise:

1. Prime numbers have structures governing their distribution among natural numbers.

2. Prime numbers are not randomly distributed as is commonly believed. Their presence or absence in the natural number series can be predicted.

3. The prime property of an integer can be understood and explained by their prime structure; this position is at variance with Harady's[3] contention that integer 317 because it is so.

TABLE 1 - Sheet 1 of 6

PRIME TABLE GENERATED WITH EQUATIONS 1 & 2

NA	NA	NB	NB	6*NA -1	6*NB + 1
Values that can be generated from EQ.1 that give compound primes	Values that cannot be generated from EQ.1 that give 6*NA - 1 Primes	Values that can be generated from EQ.2 that give compound primes	Values that cannot be generated from EQ.2 that give 6*NB + 1 Primes	Primes generated with NA of Column 2	Primes generated with NB of Column 4
Column 1	Column 2	Column 3	Column 4	Column 5	Column 6
	1		1	5	7
	2		2	11	13
	3		3	17	19
	4	4		23	
	5		5	29	31
6			6		37
	7		7	41	43
	8	8		47	
	9	9		53	
	10		10	59	61
11			11		67
	12		12	71	73
13			13		79
	14	14		83	
	15	15		89	
16			16		97
	17		17	101	103
	18		18	107	109
	19	19		113	
20		20			
21			21		127
	22	22		131	
	23		23	137	139
24		24			
	25		25	149	151
26			26		157
27			27		163

TABLE 1 - Sheet 2 of 6

PRIME TABLE GENERATED WITH EQUATIONS 1 & 2

NA	NA	NB	NB	6*NA -1	6*NB + 1
Values that can be generated from EQ.1 that give compound primes	Values that cannot be generated from EQ.1 that give 6*NA - 1 Primes	Values that can be generated from EQ.2 that give compound primes	Values that cannot be generated from EQ.2 that give 6*NB + 1 Primes	Primes generated with NA of Column 2	Primes generated with NB of Column 4
Column 1	Column 2	Column 3	Column 4	Column 5	Column 6
	28	28		167	
	29	29		173	
	30		30	179	181
31			31		
	32		32	191	193
	33		33	197	199
34		34			
35			35		211
36		36			
37			37		223
	38		38	227	229
	39	39		233	
	40		40	239	241
41			41		
	42	42		251	
	43	43		257	
	44	44		263	
	45		45	269	271
46			46		277
	47		47	281	283
48		48			
	49	49			293
50		50			
51			51		307
	52		52	311	313
	53	53		317	
54		54			

TABLE 1 - Sheet 3 of 6

PRIME TABLE GENERATED WITH EQUATIONS 1 & 2

NA	NA	NB	NB	6*NA -1	6*NB + 1
Values that can be generated from EQ.1 that give compound primes	Values that cannot be generated from EQ.1 that give 6*NA - 1 Primes	Values that can be generated from EQ.2 that give compound primes	Values that cannot be generated from EQ.2 that give 6*NB + 1 Primes	Primes generated with NA of Column 2	Primes generated with NB of Column 4
Column 1	Column 2	Column 3	Column 4	Column 5	Column 6
55			55		331
56			56		337
57		57			
	58		58	347	349
	59	59		353	
	60	60		359	
61			61		367
62			62		373
63			63		379
	64	64		383	
	65	65		389	
66			66		397
	67	67		401	
68			68		409
69		69			
	70		70	419	421
71		71			
	72		72	431	433
73			73		439
	74	74		443	
	75	75		449	
76			76		457
	77		77	461	463
	78	78		467	
79		79			
	80	80		479	
81			81		487
	82	82		491	

TABLE 1 - Sheet 4 of 6

PRIME TABLE GENERATED WITH EQUATIONS 1 & 2

NA	NA	NB	NB	6*NA -1	6*NB + 1
Values that can be generated from EQ.1 that give compound primes	Values that cannot be generated from EQ.1 that give 6*NA - 1 Primes	Values that can be generated from EQ.2 that give compound primes	Values that cannot be generated from EQ.2 that give 6*NB + 1 Primes	Primes generated with NA of Column 2	Primes generated with NB of Column 4
Column 1	Column 2	Column 3	Column 4	Column 5	Column 6
83			83		499
	84	84		503	
	85	85		509	
86			86		
	87		87	521	523
88			88		
89			89		
90			90		541
91			91		547
92		92			
	93	93		557	
	94	94		563	
	95		95	569	571
96			96		577
97		97			
	98	98		587	
	99	99		593	
	100		100	599	601
101			101		607
102			102		613
	103		103	617	619
104		104			
105			105		631
106		106			
	107		107	641	643
	108	108		647	
	109	109		653	
	110		110	659	661

TABLE 1 - Sheet 5 of 6

PRIME TABLE GENERATED WITH EQUATIONS 1 & 2

NA	NA	NB	NB	6*NA -1	6*NB + 1
Values that can be generated from EQ.1 that give compound primes	Values that cannot be generated from EQ.1 that give 6*NA - 1 Primes	Values that can be generated from EQ.2 that give compound primes	Values that cannot be generated from EQ.2 that give 6*NB + 1 Primes	Primes generated with NA of Column 2	Primes generated with NB of Column 4
Column 1	Column 2	Column 3	Column 4	Column 5	Column 6
111			111		
112				112	673
	113	113		677	
	114	114		683	
115				115	691
116			116		
	117	117		701	
118				118	709
119			119		
	120	120		719	
121				121	727
122				122	733
123				123	739
	124	124		743	
125				125	751
126				126	757
	127	127		761	
128				128	769
	129	129		773	
130			130		
131				131	787
132			132		
	133	133		797	
134			134		
	135			135	809 811
136			136		
	137			137	821 823

TABLE 1 - Sheet 6 of 6

PRIME TABLE GENERATED WITH EQUATIONS 1 & 2

NA	NA	NB	NB	6*NA -1	6*NB + 1
Values that can be generated from EQ.1 that give compound primes	Values that cannot be generated from EQ.1 that give 6*NA - 1 Primes	Values that can be generated from EQ.2 that give compound primes	Values that cannot be generated from EQ.2 that give 6*NB + 1 Primes	Primes generated with NA of Column 2	Primes generated with NB of Column 4
Column 1	Column 2	Column 3	Column 4	Column 5	Column 6
	138		138	827	829
139		139			
	140	140		839	
141		141			
142			142		853
	143		143	857	859
	144	144		863	
145		145			
146			146		877
	147		147	881	883
	148	148		887	
149		149			
150		150			
151			151		907
	152	152		911	
153			153		919
154		154			
	155	155		929	
156			156		937
	157	157		941	
	158	158		947	
	159	159		953	
160		160			
161			161		967
	162	162		971	
	163	163		977	
	164	164		983	
165			165		991
166			166		997

TABLE 2 Sheet 1 of 1

COMPARING

A: $\pi(100) - \pi(0)$ AND   B: $\pi(10,000,100) - \pi(10,000,000)$

SHOWING PRIMES AND COMPOUND PRIMES IN THESE INTERVALS

A   B

Primes 2 & 3

NA/NB	6*NA - 1	6*NB + 1	NA/NB	6*NA - 1	6*NB + 1
1	5	7	1666667	10,000,001	10,000,003
	Prime	Prime		11;909091	13;769231
2	11	13	1666668	10,000,007	10,000,009
	Prime	Prime		941;10627	23; 434783
3	17	19	1666669	10,000,013	10,000,015
	Prime	Prime		421;23753	5;2000003
4	23	25	1666670	10,000,019	10,000,021
	Prime	5;5		PRIME	97;103093
5	29	31	1666671	10,000,025	10,000,027
	Prime	Prime		5;5;7;57143	37;270271
6	35	37	1666672	10,000,031	10,000,033
	5;7	Prime		227;44053	397;25189
7	41	43	1666673	10,000,037	10,000,039
	Prime	Prime		43;313;743	7;601;2377
8	47	49	1666674	10,000,043	10,000,045
	Prime	7;7		2089;4787	5;11;11;16529
9	53	55	1666675	10,000,049	10,000,051
	Prime	5;11		47;263;809	73;136987
10	59	61	1666676	10,000,055	10,000,057
	Prime	Prime		5;13;23;6689	79;126583
11	65	67	1666677	10,000,061	10,000,063
	5;13	Prime		19;19;27701	17;588239
12	71	73	1666678	10,000,067	10,000,069
	Prime	Prime		7;7;11;18553	181;55249
13	77	79	1666679	10,000,073	10,000,075
	7;11	Prime		31;322583	5;5;269;1487
14	83	85	1666680	10,000,079	10,000,081
	Prime	5;17		PRIME	7;13;109891
15	89	91	1666681	10,000,085	10,000,087
	Prime	7;13		5;67;29851	59;169493
16	95	97	1666682	10,000,091	10,000,093
	5;19	Prime		251;39841	53;188681
			1666683	10,000,097	10,000,099
				17;588241	19;29;18149

NUMBER OF PRIMES LESS THAN OR EQUAL TO ANY NUMBER N: A CALCULATION

Calculating the number of primes less than or equal to any number N, with N being a positive integer greater than or equal to 0, has been a favorite pastime of mathematicians ever since Legendere[4] put forward the logarithmic law for the frequency of primes. This was modified by Gauss[4] and further refined by Riemann[4]. The logarithmic law was Initially put forward as a guess but was later independently proved by Hadamard and Valle-Poussin. Now known as Prime Number Theorem, it is an important landmark in the history of prime numbers. The logarithmic law and its modifications provide a good estimate of the number of primes below a positive integer N, but differ from the actual count of prime numbers.

In their article "Prime Number Races", Andrew Granville and Greg Martin[5] find no pattern in variation of the number of prime numbers generated with differing prime representation. It is important to note their study did not include (6*n ± 1) prime representation.

Calculation of Number of Primes.

Prime representations (6*NA − 1) and (6*NB + 1) dictate for a positive integer N, the total number of derived primes and compound primes of each group of NA and NB are given by Pna and Pnb per Table A

TABLE A

MOD(N,6)	NA Group	NB Group
	Pna	Pnb
0	INT(N/6)	INT(N/6) - 1
1,2,3,4.	INT(N/6)	INT(N/6)
5	INT(6) + 1	INT(N/6)

These sites are occupied by both derived primes and compound primes for any positive integer equal to or less than N. If the number of compound primes determined by equations A and B (of chapter 2) are deducted from **Pna** and **Pnb** of Table A, the number of derived primes of each group NA and NB is obtained. The total number of primes equal to or less than for a positive integer N may be written as:

π(N) = [(Pna + Pnb) − (number of compound primes)] + 2 (for primary primes 2 and 3).

$\pi(100)$ = 2*((INT(100/6)) − (compound primes in NA and NB groups) + 2.

$\pi(100)$ = 2*16 − (5 +4) + 2 = 25

For N = 1000,

$\pi(1000)$ = 2*((INT(N/6)) − (80 + 86) +2 = 168

as there are 80 compound primes in NA group and 86 compound primes in NB group.

Translating the above concept into a computer program and ensuring that a compound prime site is not counted more than once, computer outputs show $\pi(N)$ for N= 10, 10^2, 10^3, 10^4 and 10^5 are respectively 4, 25,168,1229 and 9552 in complete agreement with what is found, demonstrating that $\pi(N)$ for N =10 to 10^5, is governed by (6*n ± 1) prime number distributions for prime number greater than 3.

Table 3 such a computation along with values of Legendre, Gauss and Riemann formulas from figure 5.2 of Book of Numbers of Guy and Conway[5] . With a good software program and fast computer like super computer, it should be possible to extend this approach for N > 10^5.

TABLE 3 Sheet 1 of 1

ERRORS IN (TO THE NEAREST INTEGER) IN FORMULA OF $\pi(N)$
(BASED ON FIG.5.2 OF THE BOOK OF NUMBERS BY
JOHN H.CONWAY AND RICHARD K GUY)

NUMBER OF PRIMES UP TO N   ...... AND THE ERRORS IN

$\pi(N)$

N		LEGENDRE's FORMULA $N/\ln N - \pi(N)$	GAUSS'S GUESS	Riemann's Refinement	LJB COMPUTATION
10	4	0	2		0
10^2	25	-3	5	1	0
10^3	168	-23	10	0	0
10^4	1229	-143	17	-2	0
10^5	9592	-906	38	-5	0
10^6	78498	-6116	130	29	
10^7	664579	-44158	339	88	
10^8	5761455	-332774	754	97	
10^9	50847534	-2592592	1701	-79	
10^10	455052511	-20758029	3104	-1828	
10^11	4118054813	-169923159	11588	-2318	
10^12	37607912018	-1416705183	38263	-1476	
10^13	346065536859	-11992858452	108971	-5773	
10^14	3204941750802	-10283808636	314890	-19200	
10^15	29844570422669	-891604962452	1052619	73218	
10^16	279238341033925	-7804289844393	3214632	327052	

Computational Approach to Fremat's method of factorization of (6*n ± 1) pattern compound primes and a set of Quadratic equations to factorize compound primes or determine primality of derived primes.

Computational Approach to Fremat's method of factorization

A pattern survey of prime numbers shows primes greater than 3 fall in either 6*NA – 1 or 6*NB +1 pattern with NA and NB taking up integer values greater than 0. (6*NA – 1 pattern will be referred as A group and 6*NB + 1 pattern will be referred as B group). Further examination of these patterns shows these patterns yield primes only for certain integer values of NA and NB and yield compound primes for other integer values of NA and NB. Such compound primes have primes as factors as they are the result of multiplication of primes greater than 3. NA and NB values at which compound primes will be generated is given by

$$NA = 6*n_1*n_2 \pm (n_1 - n_2) \ldots \ldots \text{ Equation A.}$$

and

$$NB = 6*n_1*n_2 \pm (n_1 + n_2) \ldots \ldots \text{ Equation B.}$$

with n_1 and n_2 taking up positive integer values greater than 0. Formula A can generate only certain integers 6,11,13,16...etc. for NA but cannot generate integers 1,2,3,4,5,7,8,9...etc and these ungeneratable values when substituted for NA in $6*NA - 1$ representation yield 5,11,17,23,29,41,47,53 ...etc. Similarly formula B can only generate integers 4,8,9 ...etc but cannot generate integers 1,2,3,5,6,7 ...etc. and these ungeneratable values when substituted for NB in $6*NB + 1$ representation yield primes 7,13,19,31,37,43 ...etc.

Generation of Fremat formulae for $6*n \pm 1$ pattern compound prime numbers

$6*n \pm 1$ pattern compound prime numbers, as stated above, are the result of prime multiplication with such multiplications taking place as follows:

1. Prime of A group with prime of B group and vice versa

2. Prime of A group with same or another prime of A group.

3. Prime of B group with same or another prime of B group.

Fremat's formulas for the respective cases may be written as

For compound primes of A group

$$= ((3*(n_1 + n_2))^2 \pm (((3*(n_1 - n_2)) -1)^2$$

For compound primes of B group

$$= (((3*(n_1 - n_2)) -1)^2 \pm ((3*(n_1 + n_2))^2$$

Some applications of Fremat's Formulas

Bearing in mind that all compound primes are positive odd integers and if they are to be expressed as difference of two positive integers, if one of the differing positive integers is odd, the other differing integer must be even or vice versa.

Suggested Steps:

Let 2635267 be the integer.

1. Determine Group of Integer. As (2635267+1)/6 =439378, it is A group.

2. Determine odd/even chracterisic of $(n_1 - n_2)$ from equation A.

   (If remainder is 0,2 or 4 it is even and if remainder 1,3 or 5,it is odd.)

   As division of 439378 by 6 gives 4 as remainder, $(n_1 - n_2)$ is even.

3. $((3*(n_1 - n_2)) -1)^2$ will therefore have to be odd and $((3*(n_1 + n_2))^2$ will therefore have to be even to produce odd compound prime.

4. To determine $((3*(n_1 + n_2))$, take square root of of compound prime 2635267.

   = 1623.66. The first even integer greater than 1623 satisfying $((3*(n_1 + n_2))$

   Is 1626.

5. Do iterations $((3*(n_1 + n_2))^2 \pm (((3*(n_1 - n_2)) -1)^2$ starting with 1626, 1632 ...

   for $((3*(n_1 + n_2))$ till 1734 in $((3*(n_1 + n_2))^2 \pm (((3*(n_1 - n_2)) -1)^2$

   $2635267 = (1734)^2 - (763)^2$, thus giving 1734+763 =2557 and 1734 -763=1031 as factors.

Using similar arguments and using appropriate Fremat formula, the factors of

3956273 = 1439*2657

5029147 = 1811*2777

11200729 = 2953*3793

A Set of Quadratic Equations to Factorize Compound Primes or Determine Primality of

Derived Primes.

$(6*n \pm 1)$ type odd positive integers have the following characteristics.

1. Primes greater than 3 are of $(6*NA - 1)$ or $(6*NB + 1)$ type with NA and NB taking up certain positive integer values. These group are called NA and NB groups respectively.

2. Multiplication of primes of NA group with primes of NB group or vice-versa yield compound primes of $(6*NA - 1)$ type at NA values given by

$$NA = 6*n_1*n_2 \pm (n_1 - n_2)$$

and multiplication of primes of NA or NB group primes with primes of same group yield compound primes of (6*NB + 1) type at NB values given by

$$NB = 6*n_1*n_2 \pm (n_1 + n_2)$$

where n_1 and n_2 are positive integers taking up values greater than 0.

1. Values of **NA** and **NB** that **cannot** be generated by above equations when substituted in (6*NA − 1) or (6*NB + 1) representations yield primes greater than 3.

2. Therefore any (6*n ± 1) type odd positive integer will either be a prime or compound primes of (6*n ± 1) type primes as factors.

3. Digital addition of:

   (a.) The first three (6*NA − 1) type integers 5,11,17 are 5,2 and 8. This digital addition repeats itself in subsequent (6*NA − 1) type integers − primes and compound primes − due to common difference 6 in (6*NA -1) integers.

   (b.) The first three (6*NB + 1) type integers 7,13,19 are 7,4 and 1. This digital addition repeats itself in subsequent (6*NB + 1) type integers – primes and compound primes – due to common difference 6 in (6*NB +1) integers.

Factoring of (6*NA – 1) type Compound Primes

Let "X" be a (6*NA – 1) type compound prime. As "X" is produced by multiplication of NA group primes with NB group primes or vice versa, do the following operations on "X".

((X+1)/6) = Y; Divide Y by will leave a remainder "R".

Use "R" in the following equations 1 to 12. These equations contain a variable "a" with its integer value varying from "0" ... "1+ INT((SQ.RT((X ± 1)/36)))/3).

For derivation of equations 1 to 12 see Appendix

In equations 1 to 36 of Chapter 4 and Appendix:

1. The term digital addition refers to digital addition of (6*n±1)integer.

2. Multiplication of digits 5 with1...etc refers to multiplication of (6*n±1) integer with digital addition 5 with (6*n±1) with digital addition 1....

Equations for Evaluating NA and NB
DIGITAL ADDITION 5 OF "X".

Remainder "R" will be either 4 or 1.

DIGITAL ADDITION 5 OF "X", R = 4

Obtained by:

Multiplication of NA group prime with NB group prime with NA <NB.

i.e. **By multiplication of digits 5 with 1, 2 with 7, and 8 with 4.**

$$NA = \frac{-(2+6*a) \pm \sqrt{(2+6*a)^2 + 4\left(\frac{X+13}{36}+a\right)}}{2}$$

Equation 1

DIGITAL ADDITION 5 OF "X", R = 4

Multiplication of NB group prime with NA group prime with NB <NA.

i.e. **By multiplication of digits 7 with 2, 4 with 8, and 1 with 5.**

$$NB = \frac{-(4+6*a) \pm \sqrt{(4+6*a)^2 + 4(\frac{X-23}{36}-a)}}{2}$$

Equation 2.

DIGITAL ADDITION 5 OF "X", R = 1

Obtained by:

Multiplication of NA group prime with NB group prime with NA <NB.

i.e. **By multiplication of digits 5 with 1, 2 with 7, and 8 with 4.**

$$NA = \frac{-(5+6*a)\pm\sqrt{(5+6*a)^2+4(\frac{X+31}{36}+a)}}{2}$$

Equation 3.

DIGITAL ADDITION 5 OF "X", R = 1

Obtained by:

Multiplication of NB group prime with NA group prime with NB <NA.

i.e. **By multiplication of digits 7 with 2, 4 with 8, and 1 with 5.**

$$NB = \frac{-(1+6*a)\pm\sqrt{(1+6*a)^2+4(\frac{X-5}{36}-a)}}{2}$$

Equation 4.

DIGITAL ADDITION 2 OF "X".

Remainder "R" will be either 5 or 2.

DIGITAL ADDITION 2 OF "X", R = 5

Obtained by:

Multiplication of NA group prime with NB group prime with NA <NB.

i.e. **By multiplication of digits 5 with 4, 2 with 1, and 8 with 7.**

$$NA = \frac{-(1+6*a) \pm \sqrt{(1+6*a)^2 + 4\left(\frac{X+7}{36} + a\right)}}{2}$$

Equation 5

DIGITAL ADDITION 2 OF "X", R = 5

Obtained by:

Multiplication of NB group prime with NA group prime with NB <NA.

i.e. **By multiplication of digits 7 with 8, 4 with 5, and 1 with 2.**

$$NB = \frac{-(5+6*a) \pm \sqrt{(5+6*a)^2 + 4(\frac{X-29}{36} - a)}}{2}$$

Equation 6.

DIGITAL ADDITION 2 OF "X", R=2

Obtained by:

Multiplication of NA group prime with NB group prime with NA <NB.

i.e. **By multiplication of digits 5 with 4, 2 with 1, and 8 with 7.**

$$NA = \frac{-(4+6*a) \pm \sqrt{(4+6*a)^2 + 4\left(\frac{X+25}{36} + a\right)}}{2}$$

Equation 7

DIGITAL ADDITION 2 OF "X", R = 2

Obtained by:

Multiplication of NB group prime with NA group prime with NB <NA.

i.e. **By multiplication of digits 7 with 8,4 with 5, and 1 with 2.**

$$NB = \frac{-(2+6*a) \pm \sqrt{(2+6*a)^2 + 4(\frac{X-11}{36} - a)}}{2}$$

Equation 8

DIGITAL ADDITION 8 OF "X"

Remainder "R" will be either 0 or 3.

DIGITAL ADDITION 8 OF "X", R = 0

Obtained by:

Multiplication of NA group prime with NB group prime with NA <NB.

i.e. **By multiplication of digits 5 with 7, 2 with 4, and 8 with 1.**

$$NA = \frac{-(6*a) \pm \sqrt{(6*a)^2 + 4\left(\frac{X+1}{36}+a\right)}}{2}$$

Equation 9.

DIGITAL ADDITION 8 OF "X", R = 0

Obtained by:

Multiplication of NB group prime with NA group prime with NB <NA.

i.e. **By multiplication of digits 7 with 5, 4 with 2, and 1 with 8.**

$$NB = \frac{-(6a) \pm \sqrt{(6a)^2 + 4(\frac{X+1}{36}-a)}}{2}$$

Equation 10

DIGITAL ADDITION 8 OF "X", R = 3

Obtained by:

Multiplication of NA group prime with NB group prime with NA <NB.

i.e. **By multiplication of digits 5 with 7, 2 with 4, and 8 with 1.**

$$NA = \frac{-(6*a+3)\pm\sqrt{(6*a+3)^2+4\left(\frac{X+19}{36}+a\right)}}{2}$$

Equation 11

DIGITAL ADDITION 8 OF "X", R = 3

Obtained by:

Multiplication of NB group prime with NA   group prime with NB <NA.

i.e. **By multiplication of digits 7 with 5, 4 with 2, and 1 with 8.**

$$NB = \frac{-(6*a+3)\pm\sqrt{(6*a+3)^2+4\left(\frac{X-17}{36}-a\right)}}{2}$$

Equation 12.

Factoring (6*NB + 1) type compound primes.

Let "X" be a (6NB + 1) type compound prime. As "X" is produced by multiplication of primes of

NA group with primes of NA group or primes of NB group with primes of NB group, perform the

following operations on "X".

$((X - 1)/6) = Y$; Dividing Y by 6 will leave r remainder "R".

Use "R" in the following equations 13 – 36. These equations carry a variable "a" with its integer

value varying from "0" .... $1 + INT((SQ.RT((X + 1)/36))/3)$".

For derivation of Equations 13 to 36 see Appendix.

DIGITAL ADDITION 7 OF "X".
Remainder "R" will be either 4 or 1.
DIGITAL ADDITION 7 OF "X", R = 4
Obtained by:
Multiplication of NA group prime with NA group prime with $NA_1 < NA_2$.
i.e. By multiplication of digits 5 with 5, 8 with 2

Multiplication of digits 5 with 5.

$$NA_1 = \frac{-(18*a-1)\pm\sqrt{(18*a-1)^2+12\left(\frac{X-1}{12}+3*a\right)}}{6}$$ Equation 13.

DIGITAL ADDITION 7 OF "X", R = 4

Obtained by:

Multiplication of NA group prime with NA group prime with $NA_1 < NA_2$.

i.e. By multiplication of digits 5 with 5, 8 with 2.

Multiplication of digits 8 with 2.

$$NA_1 = \frac{-(18*a+5)\pm\sqrt{(18*a+5)^2+12\left(\frac{X+11}{12}+3*a\right)}}{6}$$ Equation 14

DIGITAL ADDITION 7 OF "X", R = 4

Obtained by:

Multiplication of NB group prime with NB group prime with $NB_1 < NB_2$.

i.e. By multiplication of digits 4 with 4, 7 with 1.

Multiplication of digits 4 with 4.

$$NB_1 = \frac{-(18*a+1)\pm\sqrt{(18*a+1)^2+12\left(\frac{X-1}{12}-3*a\right)}}{6}$$ Equation 15

DIGITAL ADDITION 7 OF "X", R = 4

Obtained by:

Multiplication of NB group prime with NB group prime with $NB_1 < NB_2$.

i.e. By multiplication of digits 4 with 4, 7 with 1.

Multiplication of digits 7 with 1.

$$NB_1 = \frac{-(18*a+7) \pm \sqrt{(18*a+7)^2 + 12\left(\frac{X-13}{12} - 3*a\right)}}{6}$$

Equation 16

DIGITAL ADDITION 7 OF "X", R = 1

Obtained by:

Multiplication of NA group prime with NA group prime with $NA_1 < NA_2$.

i.e. By multiplication of digits 5 with 5, 2 with 8.

Multiplication of digits 5 with 5.

$$NA_1 = \frac{-(18*a+8) \pm \sqrt{(18*a+8)^2 + 12\left(\frac{X+17}{12} + 3*a\right)}}{6}$$

Equation 17

DIGITAL ADDITION 7 OF "X", R = 1

Obtained by:

Multiplication of NA group prime with NA group prime with $NA_1 < NA_2$.

i.e. By multiplication of digits 5 with 5, 2 with 8.

Multiplication of digits 2 with 8.

$$NA_1 = \frac{-(18*a+2)\pm\sqrt{(18*a+2)^2+12\left(\frac{X+5}{12}+3*a\right)}}{6}$$ Equation 18

DIGITAL ADDITION 7 OF "X", R = 1

Obtained by:

Multiplication of NB group prime with NB group prime with $NB_1 < NB_2$.

i.e. By multiplication of digits 4 with 4, 7 with 1.

Multiplication of digits 4 with 4.

$$NB_1 = \frac{-(18*a+10)\pm\sqrt{(18*a+10)^2+12(\frac{X-19}{12}-3*a)}}{6}$$ Equation 19.

DIGITAL ADDITION 7 OF "X", R = 1

Obtained by:

Multiplication of NB group prime with NB group prime with $NB_1 < NB_2$.

i.e. By multiplication of digits 4 with 4, 7 with 1.

Multiplication of digits 7 with 1.

$$NB_1 = \frac{-(18*a+16)\pm\sqrt{(18*a+16)^2+12\left(\frac{X-31}{12}-3*a\right)}}{6}$$ Equation 20

DIGITAL ADDITION 4 OF "X".

Remainder "R" will be either 2 or 5.

DIGITAL ADDITION 4 OF "X", R = 2

Obtained by:

Multiplication of NA group prime with NA group prime with $NA_1 < NA_2$.

i.e. By multiplication of digits 2 with 2, 5 with 8.

Multiplication of digits 2 with 2.

$$NA_1 = \frac{-(18*a-1) \pm \sqrt{(18*a-1)^2 + 12\left(\frac{X-1}{12} + 3*a\right)}}{6}$$ Equation 21.

DIGITAL ADDITION 4 OF "X", R = 2

Obtained by:

Multiplication of NA group prime with NA group prime with $NA_1 < NA_2$.

i.e. By multiplication of digits 4 with 4, 5 with 8.

Multiplication of digits 5 with 8.

$$NA_1 = \frac{-(18*a+5) \pm \sqrt{(18*a+5)^2 + 12\left(\frac{X+11}{12} + 3*a\right)}}{6}$$ Equation 22.

DIGITAL ADDITION 4 OF "X", R = 2

Obtained by:

Multiplication of NB group prime with NB group prime with NB_1 < NB_2.

i.e. By multiplication of digits 7 with 7, 4 with 1.

Multiplication of digits 7 with 7.

$$NB_1 = \frac{-(18*a+1)\pm\sqrt{(18*a+1)^2+12\left(\frac{X-1}{12}-3*a\right)}}{6}$$ **Equation 23.**

DIGITAL ADDITION 4 OF "X", R = 2

Obtained by:

Multiplication of NB group prime with NB group prime with NB_1 < NB_2.

i.e. By multiplication of digits 7 with 7, 4 with 1.

Multiplication of digits 4 with 1.

$$NB_1 = \frac{-(18*a+13)\pm\sqrt{(18*a+13)^2+12\left(\frac{X-25}{12}-3*a\right)}}{6}$$ **Equation 24.**

DIGITAL ADDITION 4 OF "X", R = 5

Obtained by:

Multiplication of NA group prime with NA group prime with NA_1 < NA_2.

i.e. By multiplication of digits 2 with 2, 5 with 8.

Multiplication of digits 2 with 2.

$$NA_1 = \frac{-(18*a+8) \pm \sqrt{(18*a+8)^2 + 12\left(\frac{X+17}{12}+3*a\right)}}{6}$$

Equation 25.

DIGITAL ADDITION 4 OF "X", R = 5

Obtained by:

Multiplication of NA group prime with NA group prime with $NA_1 < NA_2$.

i.e. By multiplication of digits 4 with 4, 5 with 8.

Multiplication of digits 5 with 8.

$$NA_1 = \frac{-(18*a+14) \pm \sqrt{(18*a+14)^2 + 12\left(\frac{X+29}{12}+3*a\right)}}{6}$$

Equation 26.

DIGITAL ADDITION 4 OF "X", R = 5

Obtained by:

Multiplication of NB group prime with NB group prime with $NB_1 < NB_2$.

i.e. By multiplication of digits 7 with 7, 4 with 1.

Multiplication of digits 7 with 7.

$$NB_1 = \frac{-(18*a+10) \pm \sqrt{(18*a+10)^2 + 12\left(\frac{X-19}{12}-3*a\right)}}{6}$$

Equation 27.

DIGITAL ADDITION 4 OF "X", R = 5

Obtained by:

Multiplication of NB group prime with NB group prime with $NB_1 <NB_2$.

i.e. By multiplication of digits 7 with 7, 4 with 1.

Multiplication of digits 4 with 1.

$$NB_1 = \frac{-(18*a+4) \pm \sqrt{(18*a+4)^2 + 12\left(\frac{X-7}{12} - 3*a\right)}}{6}$$

Equation 28.

DIGITAL ADDITION 1 OF "X".

Remainder "R" will be either 0 or 3.

DIGITAL ADDITION 1 OF "X", R =0

Obtained by:

Multiplication of NA group prime with NA group prime with $NA_1 <NA_2$.

i.e. By multiplication of digits 8 with 8,5 with2.

Multiplication of digits 8 with 8.

$$NA_1 = \frac{-(18*a-1) \pm \sqrt{(18*a-1)^2 + 12\left(\frac{X-1}{12} + 3*a\right)}}{6}$$

Equation 29

DIGITAL ADDITION 1 OF "X", R = 0

Obtained by:

Multiplication of NA group prime with NA group prime with $NA_1 < NA_2$.

i.e. By multiplication of digits 8 with 8,5 with2.

Multiplication of digits 5 with 2.

$$NA_1 = \frac{-(18*a+11) \pm \sqrt{(18*a+11)^2 + 12\left(\frac{X+23}{12}+3*a\right)}}{6}$$

Equation 30

DIGITAL ADDITION 1 OF "X", R = 0

Obtained by:

Multiplication of NB group prime with NB group prime with $NB_1 < NB_2$.

i.e. By multiplication of digits 1 with 1, 4 with 7.

Multiplication of digits 1 with 1.

$$NB_1 = \frac{-(18*a+1) \pm \sqrt{(18*a+1)^2 + 12\left(\frac{X-1}{12}-3*a\right)}}{6}$$

Equation 31

DIGITAL ADDITION 1 OF "X", R = 0

Obtained by:

Multiplication of NB group prime with NB group prime with $NB_1 < NB_2$.

i.e. By multiplication of digits 1 with 1, 7 with 4.

Multiplication of digits 7 with 4.

$$NB_1 = \frac{-(18*a+13) \pm \sqrt{(18*a+13)^2 + 12\left(\frac{X-25}{12} - 3*a\right)}}{6}$$

Equation 32

DIGITAL ADDITION 1 OF "X".

Remainder "R" will be either 0 or 3.

DIGITAL ADDITION 1 OF "X", R = 3

Obtained by:

Multiplication of NA group prime with NA group prime with $NA_1 < NA_2$.

i.e. By multiplication of digits 8 with 8, 5 with 2.

Multiplication of digits 8 with 8.

$$NA_1 = \frac{-(18*a+8) \pm \sqrt{(18*a+8)^2 + 12\left(\frac{X+17}{12} + 3*a\right)}}{6}$$

Equation 33

DIGITAL ADDITION 1 OF "X", R = 3

Obtained by:

Multiplication of NA group prime with NA group prime with $NA_1 < NA_2$.

i.e. By multiplication of digits 8 with 8, 5 with 2.

Multiplication of digits 5 with 2.

$$NA_1 = \frac{-(18*a+2) \pm \sqrt{(18*a+2)^2 + 12\left(\frac{X+5}{12} + 3*a\right)}}{6}$$

Equation 34

DIGITAL ADDITION 1 OF "X", R = 3

Obtained by:

Multiplication of NB group prime with NB group prime with $NB_1 < NB_2$.

i.e. By multiplication of digits 1 with 1, 7 with 4.

Multiplication of digits 1 with 1.

$$NB_1 = \frac{-(18*a+10) \pm \sqrt{(18*a+10)^2 + 12\left(\frac{X-19}{12} - 3*a\right)}}{6}$$

Equation 35.

DIGITAL ADDITION 1 OF "X", R = 3

Obtained by:

Multiplication of NB group prime with NB group prime with $NB_1 < NB_2$.

i.e. By multiplication of digits 7 with 4, 1 with 1.

Multiplication of digits 7 with 4.

$$NB_1 = \frac{-(18*a+4) \pm \sqrt{(18*a+4)^2 + 12(\frac{X-7}{12} - 3*a)}}{6}$$

Equation 36.

Application of Equations 1 to 36 for Factoring or Primality determination of (6*n ± 1) Type Integers.

Let "X" represent a (6*n ± 1) type integer.

1. Determine the group it belongs to: NA or NB group.
2. Determine its digital addition group.
3. Determine its "R".
4. Determine applicable equations and do iterations for evaluation of "n" with "a" varying from 0...1+INT(((SQ.RT((X±1)/36))/3)).

Digital Addition	"R"	Equation	Iterations for "a"
5	4	1, 2	0...1+INT(((SQ.RT((X±1)/36))/3))
5	1	3, 4	0...1+INT(((SQ.RT((X±1)/36))/3))
2	5	5, 6	0...1+INT(((SQ.RT((X±1)/36))/3))
2	2	7, 8	0...1+INT(((SQ.RT((X±1)/36))/3))

8	0	9, 10	$0...1+INT(((SQ.RT((X\pm1)/36))/3))$
8	3	11,12	$0...1+INT(((SQ.RT((X\pm1)/36))/3))$
7	4	13,14,15,16	$0...1+INT(((SQ.RT((X\pm1)/36))/3))$
7	1	17,18,19,20	$0...1+INT(((SQ.RT((X\pm1)/36))/3))$
4	2	21,22,23,24	$0...1+INT(((SQ.RT((X\pm1)/36))/3))$
4	5	25,26,27,28	$0...1+INT(((SQ.RT((X\pm1)/36))/3))$
1	0	29,30,31,32	$0...1+INT(((SQ.RT((X\pm1)/36))/3))$
1	3	33,34,35,36	$0...1+INT(((SQ.RT((X\pm1)/36))/3))$

5. (a). If the discriminant of applicable equations produces a square integer number for a value of "a", then "X" is a composite integer and factors "n" of composite

    (6*n ± 1) integer can be evaluated.

    (b). If the discriminant of applicable equations does not produce a square integer, for all iterations of "a", then the number "X" is a prime.

A few examples of Application of Equations to "X".

(a). Let "X" be 15885131.
Digital addition 1+5+8+8+5+1+3+1 =32 =3+2 = 5. i.e an NA group integer.

Remainder "R" = 4.

Applicable Equations 1 and 2.

Number of Iterations for "a": 0 …1 + INT
$(((SQ.RT((15885131+1)/36))/3))$ =0…222.

On computing discriminant of equation 1 with "a" = 10, its square root yields 1330.

"n" of NA is therefore 634 or -696.

The factors of 15885131 are +3833 and +4177 or -3833 and -4177.

(b). Let "X" be 21019057.

Digital addition of 2+1+0+1+9+0+5+7 = 25 = 2+5 =7.

Remainder "R" = 4.

Applicable Equations 13,14,15 and 16.

Number of iterations for "a": 0...1+INT

$(((SQ.RT((21019057-1)/36))/3)) = 0...255$.

On Computing discriminant of equation 16 with "a" =39, its square root yields 4639.

"n" of NB is therefore 655 or -891.3333.

The factors of 21019057 are +3931 and +5347 or -3931 and -5347.

(c). Let "X" be 25367.

Digital addition 2+5+3+6+7 = 23 = 2+3 =5.

Remainder "R" =4.

Applicable Equations 1 and 2.

Iterations for "a": 0...8

Discriminants of equations 1 and 2 for values "a" from 0...8 do not yield a square number.

Integer 25367 is therefore a Prime Number

CHAPTER 5

COMPUTATION OF GOLDBACH PRIME PAIRS

Goldbach conjecture is a conjecture that arose out of a correspondence[1] initiated by Goldbach, a Russian mathematician working at St. Petersburg Academy of Sciences, with Leonard Euler the eminent Swiss mathematician in which he expressed any even number greater than 4 can be expressed as the sum of two prime numbers. At that time this statement was made integer 1 was considered a prime. Euler in his reply stated all even numbers greater than 2 can be expressed as sum of two prime numbers. Euler believed his statement to be a theorem but it has not only eluded proof for more than 260 years, but also came to be known in mathematical world by the name Goldbach Conjecture. This conjecture has been verified to be true for large positive even Integers as large as **10^{17}** even integers.

Prime Structures

Currently a prime number is a positive integer greater than 1 with only divisors of 1 or itself with integer 1 not being considered a prime number. As integers 2 and 3 are the first positive even and odd integers greater than 1, they meet the requirements of definition for a prime and therefore called prime numbers. Positive integers greater than 3, to be considered a prime, must be divisible by either 2 and/or 3 , therefore have to take the form $(2*3*n-1)$ or $(2*3*n+1)$ with "n" taking up positive integer values greater than 0. The prime forms of 2 and 3 are therefore different from prime forms of primes greater than 3 and these two prime groups are to be treated to be distinctive from each other hereinafter referred to primary primes and derived primes respectively. It must be noted $(6*n-1)$ and $(6*n+1)$ patterns of derived primes do not yield primes for all positive integer values of "n". At some values of "n", $(6*n \pm 1)$ patterns yield factorable integers with derived primes as factors. Such $(6*n \pm 1)$ pattern positive integers are termed compound primes i.e. $(6*n \pm 1)$ pattern positive integers consist of derived primes and compound primes.

Some Consequences of $(6*n \pm 1)$ Representation

1. $(6*n \pm 1)$ representation of positive integers consist of two streams of derived primes and compound primes arranged in arithmetic progression with a common difference of 6.

2. Addition of members of each stream with any member of the same stream or another stream

produce all even numbers greater than 8 expressed as $10 + 6*k$, $12 + 6*k$ and $14 + 6*k$ patterns with "k" taking up values greater than 0. These even number patterns were to be expressed as sum of two prime numbers, such prime number components can only come from $(6*n -1)$,$(6*n - 1)$ and $(6*n + 1)$,and $(6*n +1)$ patterns of derived primes respectively.

Prime Pair Computations: Some Basics

For a positive even number $2*N$ where "N" is a positive integer greater than 4, to be expressed as Goldbach Prime Pairs , symmetrical distribution of prime numbers with respect to "N" is a good starting point. $(6*n \pm 1)$ derived primes provide such symmetrical distribution about "N". Using symbols defined in Chapter 1, such a distribution may be schematically represented as:

Puh	Cuh

Np	

Plh	Clh

with Np taking up values 0 or 1 and Plh, Puh, Clh and Cuh taking up values 0 or greater than 0.

Prime distribution of all positive even numbers can then be arranged in following four groups for computing Goldbach Prime Pairs.

Group Distribution Number	Group Distribution Pattern	Minimum Number of Prime Pairs
1.	$Puh = 0$ $Np = 1$ $Plh = 0$	$Cuh = 0$ Np $Clh = 0$
2.	$Puh > 0$ $Np = 0$ or 1	$Cuh = 0$ $Np + Puh$

3.

$Plh > 0$ $Clh = 0$

$Puh > 0$ $Cuh > 0$

$Np = 0 \text{ or } 1$ $Np + Puh$

$Plh > 0$ $Clh = 0$

4.

$Puh > 0$	$Cuh > 0$	If $Plh > Puh$
$Np = 0$ or 1		$Np + INT((Plh*Puh^2)/(NSP^2))$
		or
		If $Puh > Plh$
$Plh > 0$	$Clh = 0$	$Np + INT((Puh*Plh^2)/(NSP^2))$

Writing positive even numbers in terms of the above distribution patterns, they can be represented as:

Even Number	Distribution Option
4 and 6	1

8.

(For positive even numbers > 10, $(6*n \pm 1)$ pattern primes are only considered)

10 + 6*k pattern	12 + 6*k pattern	14 + 6*k	
10.	------------	14	1
16,22,28,34	12,18,24	20,26	2
40,46,52,58,64,70	30,36,42,48	32,38,44,50	3
76,82 ...	54,60...	56,62 ...	4

Using actual values for Plh and Puh for small even numbers and prime number theorem for computing Plh and Puh for large even numbers, Table 4 lists computations of PP(Min) for all even numbers ≥ 4, demonstrating that there is a minimum of one pair of primes whose addition equals the even number validating Goldbach Conjecture for all even numbers ≥ 4.

TABLE 4 - 1

PRIME PAIR COMPUTATIONS

NOTES:
1. NA,NB are symbols for $6*n - 1$ group and $6*n + 1$ groups respectively.
2. FOR EVEN NUMBERS Greater THAN 8, ONLY DERIVED/COMPOUND PRIMES ARE CONSIDERED.

2*N	N	Group	Np	NS	Plh NA	Plh NB	Clh NA	Clh NB	Puh NA	Puh NB	Cuh NA	Cuh NB	PPN(mn)	PPN(A)	Notes
4	2	----	1	0	0	0	0	0	0	0	0	0	1	1 : (2+2)	
6	3	----	1	0	0	0	0	0	0	0	0	0	1	1 : (3+3)	
8	4	----	0	1	1	0	0	0	0	0	0	0	1	1 : (5+3)	2
10	5	NA	1	0	0		0		0		0		1	1 : (5+5)	
12	6	NA/NB	0	1	1	0	0	0	0	1	0	0	1	1 : (7+5)	
14	7	NB	1	0		0		0		0		0	1	1 : (7+7)	
16	8	NA	0	1	1		0		1		0		1	1 : (5+11)	
18	9	NA/NB	0	2	1	1	0	0	1	1	0	0	2	2 : (5+13,7+11)	
20	10	NB	1	1	0	1	0	0	0	1	0	0	1	1 : (7+13)	
22	11	NA	0	1	2		0		1		0		2	2 : (5+17,11+11)	
24	12	NA/NB	0	3	2	1	0	0	1	2	0	0	3	3 : (5+19,7+17,11+13)	
26	13	NB	1	1		1		0		1		0	2	2: (7+19,13+13)	
28	14	NA	0	2	2		0		2		0		2	2: (5+23,11+17)	
30	15	NA/NB	0	4	2	2	0	0	2	1	0	0	3	3: (7+23,11+19,13+17)	
32	16	NB	0	2		2		0		1		1	1	1:(13+19)	
34	17	NA	1	2	2		0		2		0		3	3:(5+29,11+23,17+17)	
36	18	NA/NB	0	5	2	2	0	0	2	1	0	0	4	4:(5+31,7+29,13+23,17+19)	
38	19	NB	1	2	3	2		0		1		1	2	2:(7+31,19+19)	
40	20	NA	0	3	3		0		2		1		2	2:(11+29,17+23)	
42	21	NA/NB	0	6	3	3	0	0	2	2	1	1	4	4:(5+37,11+31,23+19, 29+13)	

TABLE 4 - 2

PRIME PAIR COMPUTATIONS

2*N	N	Group	Np	NS	Pth NA	Pth NB	Cth NA	Cth NB	Puh NA	Puh NB	Cuh NA	Cuh NB	PPN(mn)	PPN(A)	Notes
44	22	NB	0	3	--	3	--	0	--	2	--	1	2	2:(7+37,13+31)	
46	23	NA	1	3	3	--	0	--	2	--	1	--	3	3:(5+41,17+29,23+23)	
48	24	NA/NB	0	7	4	3	0	0	2	3	1	1	5	5:(5+43,11+37,17+31, 29+19,41+7)	
50	25	NB	0	3	--	3	--	0	--	3	--	0	3	3:(7+43,13+37,19+31)	
52	26	NA	0	4	4	--	0	--	3	--	1	--	3	3:(5+47,11+41,23+29)	
54	27	NA/NB	0	8	4	3	0	1	3	3	1	1	4	5:(7+47,13+41,11+41 17+37,23+31)	
56	28	NB	0	4	--	3	--	1	--	3	--	1	1	2:(13+43,19+37)	
58	29	NA	1	4	4	--	0	--	3	--	1	--	4	4:(5+53,11+47,17+41, 29+29)	
60	30	NA/NB	0	9	5	3	0	1	3	3	1	2	4	6:(7+53,13+47,17+43, 19+41,23+37,29+31)	
62	31	NB	1	4	--	3	--	1	--	2	--	2	1	2:(19+43,31+31)	
64	32	NA	0	5	5	--	0	--	4	--	1	--	4	4:(5+59,11+43,17+47, 23+41)	
66	33	NA/NB	0	10	5	4	0	1	4	3	1	2	5	6:(5+61,7+59,13+53, 19+47,23+43,29+37)	
68	34	NB	0	5	--	4	--	1	--	3	--	2	1	2:(7+61,31+37)	
70	35	NA	0	5	5	--	0	--	4	--	1	0	4	4:(11+59,17+53,23+47, 29+41)	

NOTES:
1. NA,NB are symbols for 6*n - 1 group and 6*n + 1 groups respectively.
2. FOR EVEN NUMBERS Greater THAN 8, ONLY DERIVED/COMPOUND PRIMES ARE CONSIDERED.

TABLE 4 - 3
PRIME PAIR COMPUTATIONS

NOTES:
1. NA,NB are symbols for $6*n - 1$ group and $6*n + 1$ groups respectively.
2. FOR EVEN NUMBERS Greater THAN 8, ONLY DERIVED/COMPOUND PRIMES ARE CONSIDERED.

2*N	N	Group	Np	NS	Plh NA	Plh NB	Clh NA	Clh NB	Puh NA	Puh NB	Cuh NA	Cuh NB	PPN(mn)	PPN(A)	Notes
72	36	NA/NB	0	11	5	4	1	1	4	4	1	2	4	6;(5+67,11+61,13+59, 19+53,29+43,31+41)	
74	37	NB	1	5	---	4	---	1	---	3	---	2	2	4;(7+67,13+61,31+43, 37+37)	
76	38	NA	0	6	5	---	1	---	5	---	1	---	3	4;(5+71,17+59,23+53, 29+47)	
78	39	NA/NB	0	12	5	5	1	1	5	4	1	2	5	7;(5+73,7+71,11+67, 17+61,19+59,31+47, 37+41)	
80	40	NB	0	6	---	5	---	1	---	4	---	2	2	4;(7+73,13+67,19+61, 37+41)	
82	41	NA	1	6	5	---	1	---	4	---	2	---	3	4;(11+71,23+59,29+53, 41+41)	
84	42	NA/NB	0	13	6	5	1	1	4	5	2	2	5	8;(5+79,11+73,13+71, 17+67,23+61,31+53, 37+47,41+43)	
86	43	NB	1	6	---	5	---	1	---	4	---	2	3	4;(7+79,13+73,19+67, 43+43)	
88	44	NA	0	7	6	---	1	---	5	---	2	---	3	4;(5+83,17+71,29+59, 41+47)	
90	45	NA/NB	0	14	6	6	1	1	5	4	2	3	4	9;(7+83,11+79,17+73, 19+71,23+67,29+61, 31+59,37+53,43+47)	

TABLE 4.4
PRIME PAIR COMPUTATIONS

2*N	N	Group	Np	NS	Pth NA	Pth NB	Cih NA	Cih NB	Puh NA	Puh NB	Cuh NA	Cuh NB	PPN(mn)	PPN(A)	Notes
92	46	NB	0	7	--	6	--	1	--	4	--	3	1	3;(13+79,19+73,31+61)	
94	47	NA	1	7	6	--	1	--	5	--	2	--	4	5;(5+89,11+83,23+71, 41+53,47+47)	
96	48	NA/NB	0	15	7	6	1	1	5	4	2	4	4	7;(7+89,13+83,17+79, 23+73,29+67,37+59, 43+53)	
98	49	NB	0	7	--	6	--	1	--	4	--	3	1	3;(19+79,31+67,37+61)	
100	50	NA	0	8	7	--	1	--	5	--	3	--	2	5;(11+89,17+83,29+71, 41+59,47+53)	
102	51	NA/NB	0	16	7	6	1	2	5	5	3	3	4	8;(5+97,13+89,19+83, 23+79,29+73,31+71, 41+61,43+59)	
104	52	NB	0	8	--	6	--	2	--	5	--	3	2	4;(7+97,31+73,37+67 43+61)	
196	98	NA	0	16	12	--	4	--	10	--	6	--	4	8	
198	99	NA/NB	0	32	12	11	4	5	10	9	6	7	7	13	
200	100	NB	0	16	--	11	--	5	--	9	--	7	3	7	

NOTES:
1. NA, NB are symbols for 6*n - 1 group and 6*n + 1 groups respectively.
2. FOR EVEN NUMBERS GREATER THAN 8, ONLY DERIVED/COMPOUND PRIMES ARE CONSIDERED.

TABLE 4 - 5
PRIME PAIR COMPUTATIONS

NOTES:
1. NA,NB are symbols for 6*n - 1 group and 6*n + 1 groups respectively.
2. FOR EVEN NUMBERS GREATER THAN 8, ONLY DERIVED/COMPOUND PRIMES ARE CONSIDERED.

2*N	N	Group	Np	NS	Plh NA	Plh NB	Clh NA	Clh NB	Puh NA	Puh NB	Cuh NA	Cuh NB	PPN(mn)	PPN(A)	Notes
502	251	NA	1	41	26	--	15	--	21	--	20	--	7	14	
504	252	NA/NB	0	83	27	25	15	16	21	20	20	22	12	27	
506	253	NB	0	41	--	25	--	16	--	20	--	21	5	14	
4996	2498	NA	0	416	187	--	229	--	150	--	266	--	24	62	
4998	2499	NA/NB	0	832	187	178	229	238	150	151	266	285	47	144	
5000	2500	NB	0	416	--	178	--	238	--	151	--	265	23	76	
7498	3749	NA	0	624	261	--	363	--	218	--	406	--	31	86	
7500	3750	NA/NB	0	1249	261	259	364	365	218	209	406	416	60	206	
7502	3751	NB	0	624	--	259	--	365	--	209	--	415	29	83	
10000	5000	NA	0	833	337	--	496	--	279	--	554	--	37	127	
10002	5001	NA/NB	0	1666	337	330	496	503	279	281	554	552	75	197	
10004	5002	NB	0	833	--	330	--	503	--	281	--	552	37	99	
100000	50000	NA	0	8333	2575	--	5758	--	2231	--	6102	--	184	610	
100002	50001	NA/NB	0	16666	2575	2556	5758	5777	2231	2228	6102	6105	367	1423	
100004	50002	NB	0	8333	--	2556	--	5777	--	2228	--	6105	182	627	

TABLE 4 -- 6
PRIME PAIRS COMPUTATION FOR 2*N = 10^100

2*N	1E+100
N	5E+99
Ln(2*N)	230.258509294
Ln(N)	229.565362188
(N/Ln(N))/2	Plh	1.08901E+97
(2N/Ln(2*N))/2		2.17147E+97
(2N/Ln(2*N) - N/Ln(N))/2	Puh	1.08246E+97
INT(N/6)	NSP	8.33333E+98
PRIME PAIRS	PP	1.83746E+93

APPENDIX

Derivation of Equations 1 to 36

DIGITAL ADDITION 5 OF "X".

Remainder "R" will be either 4 or 1.

DIGITAL ADDITION 5 OF "X", R = 4

Obtained by:

Multiplication of NA group prime with NB group prime with NA <NB.

i.e. **By multiplication of digits 5 with 1, 2 with 7, and 8 with 4.**

$((6*NA) -1))*((6*NB+1) =X.$

Let NA=n, NB will then be = $(n+2 +6*a)$ where $a \geq 0$.

$((6*n-1))*(6*n +36*a +13) =X$

$36n^2+216n*a+78n-6n-36a-13=X$

$n^2+2n+6n*a -a- ((X+13)/36) =0$

$n = NA =$

$$NA = \frac{-(2+6*a) \pm \sqrt{(2+6*a)^2 + 4\left(\frac{X+13}{36}+a\right)}}{2}$$

Equation 1

DIGITAL ADDITION 5 OF "X", R = 4

Obtained by:

Multiplication of NB group prime with NA group prime with NB <NA.

i.e. **By multiplication of digits 7 with 2,4 with 8, and 1 with 5.**

$((6*NB)+1))*((6*(NA) -1) =X.$

Let NB=n, NA will then be = (n+4 +6*a) where a $\geq$ 0.

$((6*n+1))*(6*(n+4+6*a)-1)=X.$

$(6*n+1)*(6*n+36*a+23)=X$

$36n^2+216n*a+138n+6n+36a+23=X$

$n^2+4n+6n*a +a-((X-23)/36) =0$

n =NB =

$$NB = \frac{-(4+6*a)\pm\sqrt{(4+6*a)^2+4(\frac{X-23}{36}-a)}}{2}$$

Equation 2.

DIGITAL ADDITION 5 OF "X", R = 1

Obtained by:

Multiplication of NA group prime with NB group prime with NA <NB.

i.e. **By multiplication of digits 5 with 1, 2 with 7, and 8 with 4.**

$((6*NA) -1))*((6*(NB*)+1) =X.$

Let NA=n, NB will then be = (n+5 +6*a) where a $\geq$ 0.

$((6*n -1))*(6*(n+5+6*a)+1) =X.$

$((6*n-1))*(6*n +36*a +31) =X$

$36n^2+216n*a+186n-6n-36a-31=X$

$n^2+5n+6n*a -a- ((X+31)/36) =0$

$n = NA =$

$$NA = \frac{-(5+6*a)\pm\sqrt{(5+6*a)^2+4(\frac{X+31}{36}+a)}}{2}$$

Equation 3.

DIGITAL ADDITION 5 OF "X", R = 1

Obtained by:

Multiplication of NB group prime with NA group prime with NB <NA.

i.e. **By multiplication of digits 7 with 2,4 with 8, and 1 with 5.**

$((6*NB)+1))*((6*(NA*) -1) =X.$

Let NB=n, NA will then be = $(n+1+6*a)$ where a ≥ 0.

$((6*n +-1))*(6*(n+1+6*a)-1) =X.$

$((6*n+1))*(6*n +36*a +5) =X$

$36n^2+216n*a+30n+6n+36a+5=X$

$n^2+n+6n*a +a-((X-5)/36) =0$

$n = NB =$

$$NB = \frac{-(1+6*a)\pm\sqrt{(1+6*a)^2+4(\frac{X-5}{36}-a)}}{2}$$

Equation 4.

DIGTAL ADDITION 2 OF "X".

Remainder "R" will be either 5 or 2.

DIGITAL ADDITION 2 OF "X", R = 5

Obtained by:

Multiplication of NA group prime with NB group prime with NA <NB.

i.e. **By multiplication of digits 5 with 4, 2 with 1, and 8 with 7.**

$((6*NA) -1))*((6*(NB*)+1) =X.$

Let NA=n, NB will then be = (n+1 +6*a) where a ≥ 0.

$((6*n-1))*(6*n +36*a +7) =X$

$36n^2+216n*a+42n-6n-36a-7=X$

$n^2+n+6n*a -a- ((X+7)/36) =0$

n = NA =

$$NA = \frac{-(1+6*a)\pm\sqrt{(1+6*a)^2+4\left(\frac{X+7}{36}+a\right)}}{2}$$

Equation 5

DIGITAL ADDITION 2 OF "X", R = 5

Obtained by:

Multiplication of NB group prime with NA group prime with NB <NA.

i.e. **By multiplication of digits 7 with 8, 4 with 5, and 1 with 2.**

$((6*NB)+1))*((6*(NA*) -1) =X.$

Let NB=n, NA will then be = (n+5+6*a) where a ≥ 0.

((6*n +-1))*(6*(n+5+6*a)-1) =X.

((6*n+1))*(6*n +36*a +29) =X

$36n^2+216n*a+174n+6n+36a+29=X$

$n^2+5n+6n*a +a-((X-29)/36) =0$

n = NB =

$$NB = \frac{-(5+6*a) \pm \sqrt{(5+6*a)^2 + 4(\frac{X-29}{36} - a)}}{2}$$ Equation 6.

Digital Addition 2 of "X", R=2

Obtained by:

Multiplication of NA group prime with NB group prime with NA <NB.

i.e. **By multiplication of digits 5 with 4, 2 with 1, and 8 with 7.**

((6*NA) -1))*((6*(NB*)+1) =X.

Let NA=n, NB will then be = (n+4 +6*a) where a ≥ 0.

((6*n-1))*(6*n +36*a +25) =X

$36n^2+216n*a+150n-6n-36a -25=X$

$n^2+n+6n*a -4a- ((X+25)/36) =0$

n = NA =

$$NA = \frac{-(4+6*a) \pm \sqrt{(4+6*a)^2 + 4(\frac{X+25}{36} + a)}}{2}$$ Equation 7

DIGITAL ADDITION 2 OF "X", R = 2

Obtained by:

Multiplication of NB group prime with NA group prime with NB <NA.

i.e. **By multiplication of digits 7 with 8,4 with 5, and 1 with 2.**

$((6*NB)+1))*((6*(NA*) -1) =X.$

Let NB=n, NA will then be = (n+2+6*a) where a ≥ 0.

$((6*n +-1))*(6*(n+2+6*a)-1) =X.$

$((6*n+1))*(6*n +36*a +11) =X$

$36n^2+216n*a+66n+6n+36a+29=X$

$n^2+2n+6n*a +a-((X-11)/36) =0$

n=NB=

$$NB= \frac{-(2+6*a)\pm\sqrt{(2+6*a)^2+4(\frac{X-11}{36}-a)}}{2}$$ **Equation 8**

DIGITAL ADDITION 8 OF "X"

Remainder "R" will be either 0 or 3.

DIGITAL ADDITION 8 OF "X", R = 0

Obtained by:

Multiplication of NA group prime with NB group prime with NA <NB.

i.e. **By multiplication of digits 5 with 7, 2 with 4, and 8 with 1.**

$((6*NA) -1))*((6*(NB*)+1) =X.$

Let NA=n, NB will then be = (n+6*a) where a ≥ 0.

((6*n-1))*(6*n +36*a +1) =X

$36n^2+216n*a+6n-6n-36a-1=X$

$n^2+6n*a -a- ((X+1)/36) =0$

n = NA =

$$NA = \frac{-(6*a)\pm\sqrt{(6*a)^2+4\left(\frac{X+1}{36}+a\right)}}{2}$$

Equation 9.

DIGITAL ADDITION 8 OF "X", R = 0

Obtained by:

Multiplication of NB group prime with NA group prime with NB <NA.

i.e. **By multiplication of digits 7 with 5, 4 with 2, and 1 with 8.**

((6*NB) +1))*((6*(NA*) -1) =X.

Let NA=n, NB will then be = (n+6*a) where a ≥ 0.

((6*n +1))*(6*n +36*a -1) =X

$36n^2+216n*a-6n+6n-36a-1=X$

$n^2+6n*a +a- ((X+1)/36) =0$

n =NB =

$$NB = \frac{-(6a)\pm\sqrt{(6a)^2+4(\frac{X+1}{36}-a)}}{2}$$

Equation 10

DIGITAL ADDITION 8 OF "X", R = 3
Obtained by:

Multiplication of NA group prime with NB group prime with NA <NB.

i.e. By multiplication of digits **5 with 7, 2 with 4, and 8 with 1.**

$((6*NA) -1))*((6*(NB)+1) =X.$

Let NA=n, NB will then be = $(n+3+6*a)$ where a ≥ 0.

$((6*n-1))*(6*n +36*a +19) =X$

$36n^2+216n*a+114n-6n-36a-19=X$

$n^2+6n*a +3n -a- ((X+19)/36) =0$

n = NA =

$$NA = \frac{-(6*a+3)\pm\sqrt{(6*a+3)^2+4\left(\frac{X+19}{36}+a\right)}}{2}$$

Equation 11

DIGITAL ADDITION 8 OF "X", R = 3
Obtained by:

Multiplication of NB group prime with NA group prime with NB <NA.

i.e. **By multiplication of digits 7 with 5, 4 with 2, and 1 with 8.**

$((6*NB) +1))*((6*(NA)-1) =X.$

Let NB=n, NA will then be = $(n+3+6*a)$ where a ≥ 0.

$((6*n+1))*(6*n +36*a +17) =X$

$36n^2+216n*a+102n+6n+36a+17=X$

$n^2+6n*a +3n +a- ((X-17)/36) =0$

$n = NB =$

$$NB = \frac{-(6*a+3)\pm\sqrt{(6*a+3)^2+4\left(\frac{X-17}{36}-a\right)}}{2}$$

Equation 12.

Factoring of (6*NB+ 1) type of composites.

Let "X" be a (6*NB + 1) type composite. "X" is prod

uced by multiplication of primes of

NA group with primes of NA group or primes of NB group with primes of NB group.

Do the following operation on "X".

$((X-1)/6) = Y$; Dividing Y by 6 will leave a remainder "R".

DIGITAL ADDITION 7 OF "X".
Remainder "R" will be either 4 or 1.
DIGITAL ADDITION 7 OF "X", R = 4
Multiplication of NA group prime with NA group prime with $NA_1 < NA_2$.
i.e. By multiplication of digits 5 with 5, 8 with 2.

Multiplication of digits 5 with 5.
$((6*NA_1) -1))*((6*NA_2) -1) = X.$
Let $NA_1 = n$, NA_2 will then be $= (n+6*a)$ where a ≥ 0.
$((6*n-1))*(6*n +36*a -1) = X$
$36n^2 + 216n*a - 6n - 6n - 36a + 1 = X$
$3n^2 - n + 18n*a - 3a - ((X -1)/12) = 0$
$n = NA_1 =$

$$NA_1 = \frac{-(18*a-1)\pm\sqrt{(18*a-1)^2+12\left(\frac{X-1}{12}+3*a\right)}}{6}$$

Equation 13.

DIGITAL ADDITION 7 OF "X", R = 4

Multiplication of NA group prime with NA group prime with $NA_1 < NA_2$.

i.e. By multiplication of digits 5 with 5, 8 with 2.

Multiplication of digits 8 with 2.

$((6*NA_1) -1))*((6*NA_2) -1) =X.$

Let $NA_1 = n$, NA_2 will then be $= (n+2+6*a)$ where $a \geq 0$.

$((6*n-1))*(6*n +36*a +11) =X$

$36n^2+216n*a +66n-6n-36a -11=X$

$3n^2+5n+18n*a -3a- ((X+11)/12) =0$

$n = NA_1 =$

$$NA_1 = \frac{-(18*a+5)\pm\sqrt{(18*a+5)^2+12\left(\frac{X+11}{12}+3*a\right)}}{6}$$

Equation 14

DIGITAL ADDITION 7 OF "X", R = 4

Multiplication of NB group prime with NB group prime with $NB_1 < NB_2$.

i.e. By multiplication of digits 4 with 4, 7 with 1.

Multiplication of digits 4 with 4.

$((6*NB_1) +1))*((6*NB_2) +1) =X.$

Let $NB_1 = n$, NB_2 will then be = $(n+6*a)$ where $a \geq 0$.

$((6*n +1))*(6*n +36*a +1) = X$

$36n^2 + 216n*a + 6n + 6n + 36a + 1 = X$

$3n^2 + n + 18n*a + 3a - ((X-1)/12) = 0$

$n = NB_1 = $

$$NB_1 = \frac{-(18*a+1) \pm \sqrt{(18*a+1)^2 + 12\left(\frac{X-1}{12} - 3*a\right)}}{6}$$

Equation 15

DIGITAL ADDITION 7 OF "X", R = 4

Multiplication of NB group prime with NB group prime with $NB_1 < NB_2$.

i.e. By multiplication of digits 4 with 4, 7 with 1.

Multiplication of digits 7 with 1.

$((6*NB_1) +1))*((6*NB_2) +1) = X$.

Let $NB_1 = n$, NB_2 will then be = $(n+2+6*a)$ where $a \geq 0$.

$((6*n +1))*(6*n +36*a +13) = X$

$36n^2 + 216n*a + 78n + 6n + 36a + 13 = X$

$3n^2 + 7n + 18n*a + 3a - ((X-13)/12) = 0$

$n = NB_1 = $

$$NB_1 = \frac{-(18*a+7) \pm \sqrt{(18*a+7)^2 + 12\left(\frac{X-13}{12} - 3*a\right)}}{6}$$

Equation 16

DIGITAL ADDITION 7 OF "X", R = 1

Multiplication of NA group prime with NA group prime with $NA_1 < NA_2$.

i.e. By multiplication of digits 5 with 5, 2 with 8.

Multiplication of digits 5 with 5.

$((6*NA_1) -1))*((6*NA_2) -1) = X$.

Let $NA_1 = n$, NA_2 will then be = $(n+3+6*a)$ where $a \geq 0$.

$((6*n-1))*(6*n +36*a +17) = X$

$36n^2 + 216n*a +102n - 6n - 36a -17 = X$

$3n^2 + 8n + 18n*a - 3a - ((X +17)/12) = 0$

$n = NA_1 =$

$$NA_1 = \frac{-(18*a+8) \pm \sqrt{(18*a+8)^2 + 12\left(\frac{X+17}{12} + 3*a\right)}}{6}$$

Equation 17

DIGITAL ADDITION 7 OF "X", R = 1

Multiplication of NA group prime with NA group prime with $NA_1 < NA_2$.

i.e. By multiplication of digits 5 with 5, 2 with 8.

Multiplication of digits 2 with 8.

$((6*NA_1) -1))*((6*NA_2) -1) = X$.

Let $NA_1 = n$, NA_2 will then be = $(n+1+6*a)$ where $a \geq 0$.

$((6*n-1))*(6*n +36*a +5) = X$

$36n^2 + 216n*a +30n - 6n - 36a -5 = X$

$$3n^2+2n+18n*a -3a- ((X+5)/12) =0$$
$$n = NA_1 =$$

$$NA_1 = \frac{-(18*a+2)\pm\sqrt{(18*a+2)^2+12\left(\frac{X+5}{12}+3*a\right)}}{6}$$ **Equation 18**

DIGITAL ADDITION 7 OF "X", R = 1

Multiplication of NB group prime with NB group prime with $NB_1 < NB_2$.

i.e. By multiplication of digits 4 with 4, 7 with 1.

Multiplication of digits 4 with 4.

$((6*NB_1) +1))*((6*NB_2) +1) =X$.

Let $NB_1=n$, NB_2 will then be $= (n+3+6*a)$ where $a \geq 0$.

$((6*n +1))*(6*n +36*a +19) =X$

$36n^2+216n*a +114n+6n+36a +19=X$

$3n^2+10n+18n*a +3a- ((X -19)/12) =0$

$n = NB_1 =$

$$NB_1 = \frac{-(18*a+10)\pm\sqrt{(18*a+10)^2+12(\frac{X-19}{12}-3*a)}}{6}$$ **Equation 19.**

DIGITAL ADDITION 7 OF "X", R = 1

Multiplication of NB group prime with NB group prime with $NB_1 < NB_2$.

i.e. By multiplication of digits 4 with 4, 7 with 1.

Multiplication of digits 7 with 1.

$((6*NB_1) +1))*((6*NB_2) +1) =X.$

Let $NB_1=n$, NB_2 will then be $= (n+5+6*a)$ where $a \geq 0$.

$((6*n +1))*(6*n +36*a +31) =X$

$36n^2+216n*a +186n+6n +36a +13=X$

$3n^2+16n+18n*a +3a- ((X-13)/12) =0$

$n = NB_1 =$

$$NB_1 = \frac{-(18*a+16)\pm\sqrt{(18*a+16)^2+12\left(\frac{X-31}{12}-3*a\right)}}{6}$$

Equation 20

DIGITAL ADDITION 4 OF "X".

Remainder "R" will be either 2 or 5.

DIGITAL ADDITION 4 OF "X", R = 2

Multiplication of NA group prime with NA group prime with $NA_1 <NA_2$.

i.e. By multiplication of digits 2 with 2, 5 with 8.

Multiplication of digits 2 with 2.

$((6*NA_1) -1))*((6*NA_2) -1) =X.$

Let $NA_1=n$, NA_2 will then be $= (n+6*a)$ where $a \geq 0$.

$((6*n-1))*(6*n +36*a -1) =X$

$36n^2+216n*a-6n-6n-36a +1=X$

$3n^2-n+18n*a -3a- ((X -1)/12) =0$

$n = NA_1 =$

$$NA_1 = \frac{-(18*a-1)\pm\sqrt{(18*a-1)^2+12\left(\frac{X-1}{12}+3*a\right)}}{6}$$

Equation 21.

DIGITAL ADDITION 4 OF "X", R = 2

Multiplication of NA group prime with NA group prime with $NA_1 < NA_2$.

i.e. By multiplication of digits 4 with 4, 5 with 8.

Multiplication of digits 5 with 8.

$((6*NA_1) -1))*((6*NA_2) -1) = X$.

Let $NA_1 = n$, NA_2 will then be $= (n+2+6*a)$ where $a \geq 0$.

$((6*n-1))*(6*n +36*a +11) = X$

$36n^2 + 216n*a + 66n - 6n - 36a - 11 = X$

$3n^2 + 5n + 18n*a - 3a - ((X+11)/12) = 0$

$n = NA_1 =$

$$NA_1 = \frac{-(18*a+5) \pm \sqrt{(18*a+5)^2 + 12\left(\frac{X+11}{12} + 3*a\right)}}{6}$$

Equation 22.

DIGITAL ADDITION 4 OF "X", R = 2

Multiplication of NB group prime with NB group prime with $NB_1 < NB_2$.

i.e. By multiplication of digits 7 with 7, 4 with 1.

Multiplication of digits 7 with 7.

$((6*NB_1) +1))*((6*NB_2) +1) = X$.

Let $NB_1 = n$, NB_2 will then be $= (n+6*a)$ where $a \geq 0$.

$((6*n +1))*(6*n +36*a +1) = X$

$36n^2 + 216n*a + 6n + 6n + 36a + 1 = X$

$3n^2 + n + 18n*a + 3a - ((X -1)/12) = 0$

$n = NB_1 =$

$$NB_1 = \frac{-(18*a+1)\pm\sqrt{(18*a+1)^2+12\left(\frac{X-1}{12}-3*a\right)}}{6}$$

Equation 23.

DIGITAL ADDITION 4 OF "X", R = 2

Multiplication of NB group prime with NB group prime with $NB_1 < NB_2$.

i.e. By multiplication of digits 7 with 7, 4 with 1.

Multiplication of digits 4 with 1.

$((6*NB_1)+1))*((6*NB_2)+1) = X$.

Let $NB_1 = n$, NB_2 will then be $= (n+4+6*a)$ where $a \geq 0$.

$((6*n+1))*(6*n+36*a+25) = X$

$36n^2+216n*a+150n+6n+36a+25 = X$

$3n^2+13n+18n*a+3a- ((X-25)/12) = 0$

$n = NB_1 =$

$$NB_1 = \frac{-(18*a+13)\pm\sqrt{(18*a+13)^2+12\left(\frac{X-25}{12}-3*a\right)}}{6}$$

Equation 24.

DIGITAL ADDITION 4 OF "X", R = 5

Multiplication of NA group prime with NA group prime with $NA_1 < NA_2$.

i.e. By multiplication of digits 2 with 2, 5 with 8.

Multiplication of digits 2 with 2.

$((6*NA_1) -1))*((6*NA_2) -1) =X$.

Let $NA_1=n$, NA_2 will then be $= (n+3+6*a)$ where $a \geq 0$.

$((6*n-1))*(6*n +36*a +17) =X$

$36n^2+216n*a +102n-6n-36a -17=X$

$3n^2+8n+18n*a -3a- ((X +17)/12) =0$

$n = NA_1 =$

$$NA_1 = \frac{-(18*a+8) \pm \sqrt{(18*a+8)^2+12\left(\frac{X+17}{12}+3*a\right)}}{6}$$

Equation 25.

DIGITAL ADDITION 4 OF "X", R = 5

Multiplication of NA group prime with NA group prime with $NA_1 <NA_2$.

i.e. By multiplication of digits 4 with 4, 5 with 8.

Multiplication of digits 5 with 8.

$((6*NA_1) -1))*((6*NA_2) -1) =X$.

Let $NA_1=n$, NA_2 will then be $= (n+5+6*a)$ where $a \geq 0$.

$((6*n-1))*(6*n +36*a +29) =X$

$36n^2+216n*a +174n-6n-36a -29=X$

$3n^2+14n+18n*a -3a- ((X+29)/12) =0$

$n = NA_1 =$

$$NA_1 = \frac{-(18*a+14) \pm \sqrt{(18*a+14)^2+12\left(\frac{X+29}{12}+3*a\right)}}{6}$$

Equation 26.

DIGITAL ADDITION 4 OF "X", R = 5

Multiplication of NB group prime with NB group prime with $NB_1 <NB_2$.

i.e. By multiplication of digits 7 with 7, 4 with 1.

Multiplication of digits 7 with 7.

$((6*NB_1) +1))*((6*NB_2) +1) =X$.

Let $NB_1=n$, NB_2 will then be $= (n+3+6*a)$ where $a \geq 0$.

$((6*n +1))*(6*n +36*a +19) =X$

$36n^2+216n*a +114n+6n+36a +19=X$

$3n^2+10n+18n*a +3a- ((X -19)/12) =0$

$$NB_1 = \frac{-(18*a+10)\pm\sqrt{(18*a+10)^2+12\left(\frac{X-19}{12}-3*a\right)}}{6}$$

Equation 27.

DIGITAL ADDITION 4 OF "X", R = 5

Multiplication of NB group prime with NB group prime with $NB_1 <NB_2$.

i.e. By multiplication of digits 7 with 7, 4 with 1.

Multiplication of digits 4 with 1.

$((6*NB_1) +1))*((6*NB_2) +1) =X$.

Let $NB_1=n$, NB_2 will then be $= (n+1+6*a)$ where $a \geq 0$.

$((6*n +1))*(6*n +36*a +7) =X$

$36n^2+216n*a +42n+6n +36a +7=X$

$3n^2+8n+18n*a +3a- ((X-7)/12) =0$

$n = NB_1 =$

$$NB_1 = \frac{-(18*a+4)\pm\sqrt{(18*a+4)^2+12\left(\frac{X-7}{12}-3*a\right)}}{6}$$

Equation 28.

DIGITAL ADDITION 1 OF "X".

Remainder "R" will be either 0 or 3.

DIGITAL ADDITION 1 OF "X", R =0

Multiplication of NA group prime with NA group prime with $NA_1 < NA_2$.

i.e. By multiplication of digits 8 with 8,5 with2.

Multiplication of digits 8 with 8.

$((6*NA_1) -1))*((6*NA_2) -1) =X$.

Let $NA_1=n$, NA_2 will then be $= (n+6*a)$ where a ≥ 0.

$((6*n-1))*(6*n +36*a -1) =X$

$36n^2+216n*a-6n-6n-36a +1=X$

$3n^2-n+18n*a -3a- ((X -1)/12) =0$

$n = NA_1 =$

$$NA_1 = \frac{-(18*a-1)\pm\sqrt{(18*a-1)^2+12\left(\frac{X-1}{12}+3*a\right)}}{6}$$

Equation 29

DIGITAL ADDITION 1 OF "X", R = 0

Multiplication of NA group prime with NA group prime with $NA_1 < NA_2$.

i.e. By multiplication of digits 8 with 8,5 with2.

Multiplication of digits 5 with 2.

$((6*NA_1) -1))*((6*NA_2) -1) =X.$

Let $NA_1=n$, NA_2 will then be = $(n+4+6*a)$ where $a \geq 0$.

$((6*n-1))*(6*n +36*a +23) =X$

$36n^2+216n*a +138n-6n-36a -23=X$

$3n^2+11n+18n*a -3a- ((X+23)/12) =0$

$n = NA_1 =$

$$NA_1 = \frac{-(18*a+11)\pm\sqrt{(18*a+11)^2+12\left(\frac{X+23}{12}+3*a\right)}}{6}$$

Equation 30

DIGITAL ADDITION 1 OF "X", R = 0

Multiplication of NB group prime with NB group prime with $NB_1 <NB_2$.

i.e. By multiplication of digits 1 with 1, 4 with 7.

Multiplication of digits 1 with 1.

$((6*NB_1) +1))*((6*NB_2) +1) =X.$

Let $NB_1=n$, NB_2 will then be = $(n+6*a)$ where $a \geq 0$.

$((6*n +1))*(6*n +36*a +1) =X$

$36n^2+216n*a +6n+6n+36a +1=X$

$3n^2+n+18n*a +3a- ((X -1)/12) =0$

$$NB_1 = \frac{-(18*a+1)\pm\sqrt{(18*a+1)^2+12\left(\frac{X-1}{12}-3*a\right)}}{6}$$

Equation 31

DIGITAL ADDITION 1 OF "X", R = 0

Multiplication of NB group prime with NB group prime with $NB_1 < NB_2$.

i.e. By multiplication of digits 1 with 1, 7 with 4.

Multiplication of digits 7 with 4.

$((6*NB_1) +1))*((6*NB_2) +1) = X$.

Let $NB_1 = n$, NB_2 will then be $= (n+4+6*a)$ where $a \geq 0$.

$((6*n +1))*(6*n +36*a +25) = X$

$36n^2 + 216n*a + 150n + 6n + 36a + 25 = X$

$3n^2 + 13n + 18n*a + 3a - ((X-25)/12) = 0$

$n = NB_1 =$

$$NB_1 = \frac{-(18*a+13) \pm \sqrt{(18*a+13)^2 + 12\left(\frac{X-25}{12} - 3*a\right)}}{6}$$

Equation 32

DIGITAL ADDITION 1 OF "X".

Remainder "R" will be either 0 or 3.

DIGITAL ADDITION 1 OF "X", R = 3

Multiplication of NA group prime with NA group prime with $NA_1 < NA_2$.

i.e. By multiplication of digits 8 with 8, 5 with2.

Multiplication of digits 8 with 8.

$((6*NA_1) -1))*((6*NA_2) -1) = X$.

Let $NA_1 = n$, NA_2 will then be $= (n+3+6*a)$ where $a \geq 0$.

$((6*n-1))*(6*n +36*a +17) = X$

$36n^2+216n*a +102n-6n-36a -17=X$

$3n^2+8n+18n*a -3a- ((X +17)/12) =0$

$n = NA_1 =$

$$NA_1 = \frac{-(18*a+8)\pm\sqrt{(18*a+8)^2 +12\left(\frac{X+17}{12}+3*a\right)}}{6}$$ **Equation 33**

DIGITAL ADDITION 1 OF "X", R = 3

Multiplication of NA group prime with NA group prime with $NA_1 < NA_2$.

i.e. By multiplication of digits 8 with 8, 5 with2.

Multiplication of digits 5 with 2.

$((6*NA_1) -1))*((6*NA_2) -1) = X.$

Let $NA_1=n$, NA_2 will then be $= (n+1+6*a)$ where $a \geq 0$.

$((6*n-1))*(6*n +36*a +5) =X$

$36n^2+216n*a +30n-6n-36a -5=X$

$3n^2+2n+18n*a -3a- ((X+5)/12) =0$

$n = NA_1 =$

$$NA_1 = \frac{-(18*a+2)\pm\sqrt{(18*a+2)^2 +12\left(\frac{X+5}{12}+3*a\right)}}{6}$$ **Equation 34**

DIGITAL ADDITION 1 OF "X", R = 3

Multiplication of NB group prime with NB group prime with $NB_1 < NB_2$.

i.e. By multiplication of digits 1 with 1, 7 with 4.

Multiplication of digits 1 with 1.

$((6*NB_1) +1))*((6*NB_2) +1) = X$.

Let $NB_1 = n$, NB_2 will then be $= (n+3+6*a)$ where $a \geq 0$.

$((6*n +1))*(6*n +36*a +19) = X$

$36n^2 +216n*a +114n +6n +36a +19 = X$

$3n^2 +10n +18n*a +3a - ((X -19)/12) = 0$

$n = NB_1 =$

$$NB_1 = \frac{-(18*a+10) \pm \sqrt{(18*a+10)^2 +12\left(\frac{X-19}{12} -3*a\right)}}{6}$$

Equation 35.

DIGITAL ADDITION 1 OF "X", R = 3

Multiplication of NB group prime with NB group prime with $NB_1 < NB_2$.

i.e. By multiplication of digits 7 with 7, 4 with 1.

Multiplication of digits 4 with 1.

$((6*NB_1) +1))*((6*NB_2) +1) = X$.

Let $NB_1 = n$, NB_2 will then be $= (n+1+6*a)$ where $a \geq 0$.

$((6*n +1))*(6*n +36*a +7) = X$

$36n^2+216n*a +42n+6n +36a +7=X$

$3n^2+8n+18n*a +3a- ((X-7)/12) =0$

$n = NB_1 =$

$$NB_1 = \frac{-(18*a+4)\pm\sqrt{(18*a+4)^2+12(\frac{X-7}{12}-3*a)}}{6}$$

Equation 36.

SUMMARY

1. Rules of multiplication and division dictate why some natural numbers meet prime number definition and others do not. For the same reason there is no simple, straightforward rule or equation for generating all prime numbers.

2. Following criteria govern generation of all prime numbers.

    a. Positive integers 2 and 3 meet prime definition because of their position in natural number sequence.

    b. Positive integers greater than 3 meet prime definition in two groups due to multiplication and division rules. The two groups are $(6*NA - 1)$ and $(6*NB + 1)$ groups, for simplicity termed NA and NB groups. The criterion for prime generation in the two groups are;

$$NA \neq 6*n_1*n_2 \pm (n_1 - n_2)$$

and

$$NB \neq 6*n_1*n_2 \pm (n_1 + n_2)$$

3. All primes are generated by steps outlined in 2(a) and 2(b) above.

4. Based on above criteria, natural numbers are grouped in seven groups.

5. The two prime groups are used to factoring compound primes using computational approach to Fremat's method of factorization and to generate a Set of quadratic equations to factorize Compound Primes or determine primality of derived primes.

6. The two prime groups are used to generate prime distribution patterns to compute prime pairs for even numbers 4 which confirm Goldbach Conjecture.

NOTES

1. Oystein Ore, Number Theory and its History
   (New York: Dover Publications, 1988)
2. C. Caldwell, Prime Pages, http://primes.utm.edu
3. G. H. Hardy, A Mathematician's Apology
   (Cambridge: Cambridge University Press, 19..)
4. J. H. Conway and R. K. Guy, The Book of Numbers
   (Copernicus, 1995)
5. Andrew Granville and G. Martin, "Prime Number Races",
   American Mathematical Monthly 113, no.1 (2006): pp 1-33.
6. R. Crandall and C. Pomerance, Prime Numbers: A Computational Perspective
   (New York: Springer Verlag, 2001)

INDEX

www.ingramcontent.com/pod-product-compliance
Lightning Source LLC
Chambersburg PA
CBHW071231170526
45165CB00003B/1073